sl Lemon

Lemons! Lemons! Lemons!

Lemons! Lemons! Lemons!

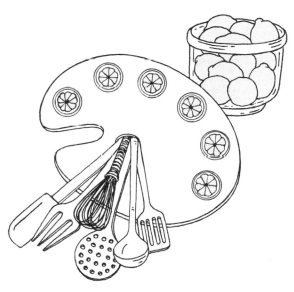

SARAH SCHULTE AND LALITTE SCOTT

VIKING

VIKING
Published by the Penguin Group
Penguin Books USA Inc., 375 Hudson Street,
New York, New York 10014, U.S.A.
Penguin Books Ltd, 27 Wrights Lane, London W8 5TZ, England
Penguin Books Australia Ltd, Ringwood, Victoria, Australia
Penguin Books Canada Ltd, 10 Alcorn Avenue,
Toronto, Ontario, Canada M4V 3B2
Penguin Books (N.Z.) Ltd, 182–190 Wairau Road,
Auckland 10, New Zealand

Penguin Books Ltd, Registered Offices:
Harmondsworth, Middlesex, England

First published in 1994 by Viking Penguin,
a division of Penguin Books USA Inc.

1   3   5   7   9   10   8   6   4   2

LIBRARY OF CONGRESS CATALOGING IN PUBLICATION DATA
Schulte, Sarah.
Lemons, lemons, lemons / Sarah Schulte and Lalitte Scott.
p.   cm.
Includes index.
ISBN 0–670–85408–5
1. Cookery (Lemons)   I. Scott, Lalitte.   II. Title.
TX813.L4S34   1994
641.6'4334—dc20   93–50698

Printed in the United States of America
Set in Bembo
Designed by Ann Gold

Contents

Lemons! Lemons! Lemons!

Lemon Notes

The lemon (*Citrus limon*), one of the most versatile and popular fruits on earth, was first mentioned in literature in A.D. 960. It originated in Burma, and from there was brought to the Middle East and eventually to Europe. Christopher Columbus introduced the lemon to Haiti, but it was not widely cultivated in America until the 1880s. Today, California, Florida, and Arizona grow half of the world's lemons, with most of the remainder coming from Italy.

There are fewer varieties of lemons than of most other citrus fruits. Eureka, a small to medium elliptical lemon, is grown in many coastal areas of California. Lisbon grows well in the warmer parts of California, and is similar to Eureka in shape and typical lemon flavor. Villafranca is a variety that has adapted well to the warmer, humid climate of Florida.

Most lemon trees require semitropical climates, growing best with hot, dry summers and cool winters. They bear fruit all year long, sometimes up to 1,000 pounds per tree. Lemons are picked while still green, and are ripened in a cool, dark place with good ventilation. If refrigerated, they remain fresh for weeks. At room temperature, however, they yield more juice than when they are cold. For maximum juice, it is best to roll the lemon before cutting. When the rind (or zest) is needed, grate the lemon before cutting and use the yellow skin only, as the white pith is bitter.

Lemons can be fragrant and medicinal and sweet or sour. The recipes here call for fresh lemons and fresh lemon juice.

When a recipe calls for mayonnaise, see page 42 or use Hellmann's Regular. Unsalted butter or unsalted margarine should be used unless otherwise specified. Chicken broth can be homemade or canned. To peel a tomato, drop it into boiling water 10 to 15 seconds and remove the skin with a fork. Scoop the seeds out with a demitasse spoon.

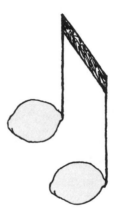

Hors d'Oeuvres and First Courses

Guacamole

1 ripe avocado, pitted and peeled

½ clove garlic, peeled and minced

½ teaspoon hot red pepper flakes

Grated rind of 1 lemon

2 tablespoons lemon juice

3 tablespoons olive oil

1 medium tomato, skinned, seeded, and diced (see Lemon Notes, page 1)

Salt to taste

Using a potato masher or a fork, mash the avocado until it is almost smooth. Add all the remaining ingredients and mix well. Serve as a dip with tortilla chips or crudités.

Yield: 8 servings.

Zucchini Spread

1 cup coarsely grated unpeeled zucchini

1 cup grated cheddar cheese

¾ cup mayonnaise (see Lemon Notes, page 1)

3 teaspoons lemon juice

2 scallions, chopped

½ cup pitted and chopped black olives, fresh or canned

½ cup coarsely chopped walnuts (optional)

Combine all the ingredients in a bowl and mix well with a fork. Cover and refrigerate overnight. Transfer to a serving dish and serve with assorted crackers or crudités.

Yield: 20 to 30 servings.

Artichoke Dip

2 14-ounce cans artichoke hearts, drained and coarsely chopped

1 cup grated Parmesan cheese

1 cup mayonnaise (see Lemon Notes, page 1)

½ cup lemon juice

1 small onion, finely chopped

Preheat the oven to 350° F.

Mix the ingredients well with a fork, and put into a round casserole dish 7 by 2½ inches. Bake 20 minutes, or until slightly browned. Serve as a dip or a spread for crackers.

Yield: 16 to 20 servings.

Salmon Mousse

8 ounces canned salmon or 1 cup leftover fresh smoked salmon

1 small onion, finely chopped

½ cup boiling water

2 envelopes unflavored gelatin

1 teaspoon chopped fresh dill

½ cup mayonnaise (see Lemon Notes, page 1)

2 tablespoons lemon juice

¼ teaspoon paprika

½ cup heavy cream

2 large cucumbers, unpeeled (optional)

Using a blender or a food processor, blend the first eight ingredients until smooth. Gradually add the cream to the mixture. Pour into a greased 2-cup mold and refrigerate overnight. Slice the cucumbers ¼ inch thick. Spread the mousse on cucumber slices or on crackers.

Yield: 16 to 20 servings.

Lemon and Egg Soup

(Avgolemono)

3   cups chicken broth
3   tablespoons long-grain white
    rice, uncooked
2   large eggs
3   tablespoons lemon juice or
    more to taste

Salt and freshly ground black
    pepper to taste
Grated rind of 1 lemon
2   tablespoons chopped fresh
    parsley

Bring the chicken broth to a boil in a large saucepan and add the rice. Cover, reduce the heat, and simmer 15 minutes. Remove from the heat.

Beat the eggs, with a rotary or electric mixer, until frothy. Continue beating while slowly adding 3 tablespoons lemon juice. Add the egg and lemon mixture to the soup a few drops at a time in order to prevent curdling; stir continuously. Season with salt and pepper, adding 1 or 2 more teaspoons lemon juice if needed for tartness. (The soup may also be served chilled. Since it may curdle if reheated, use a double boiler and stir constantly over low heat.) Sprinkle with the lemon rind and parsley right before serving.

Yield: 4 servings.

Tomato Broth

1   11-ounce can V-8 juice or
    tomato juice
1   cup chicken broth
¼   cup lemon juice
Grated rind of 1 lemon

1   stalk celery, including leaves,
    finely chopped
1   tablespoon chopped fresh
    parsley

In a medium saucepan, simmer the first four ingredients 8 minutes. Add the celery and parsley and simmer 5 more minutes. Serve hot or cold.

Yield: 3 servings.

Broccoli Soup

2 large stalks broccoli, including florets, peeled and chopped

2 medium potatoes, unpeeled, chopped

1 clove garlic, peeled and minced

1 small onion, chopped

1 small yellow squash, unpeeled, chopped

½ lemon, chopped, rind and all (with seeds removed)

5 cups chicken broth

1 teaspoon curry powder

Put all the ingredients in a large stockpot and cook over medium heat 25 minutes. Cool slightly, then place the vegetables in a blender or food processor, in batches of 2 cups at a time, adding enough broth to cover. Blend only until the mixture has a chunky consistency, being careful not to puree. Return to the pot with the remaining broth. Serve hot or cold.

Yield: 6 servings.

Mexican Soup

4   cups chicken broth
2   tablespoons long-grain white
    rice, uncooked
2   small sprigs cilantro, finely
    chopped
½   teaspoon jalapeño pepper,
    minced

2   teaspoons finely chopped
    onion
2   teaspoons lemon juice
2   tablespoons chopped avocado
4   thin slices of lemon for garnish

Bring the chicken broth to a boil in a large saucepan and add the rice. Cook, covered, 15 minutes over low heat. Add the remaining ingredients, except the avocado, to the soup and bring to a boil. Remove from the heat and add the avocado. The soup may be served hot or cold. Float a lemon slice on top of each serving.

Yield: 4 servings.

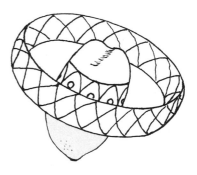

Lemon Pasta

1½ cups light cream
1 pound linguine
½ cup (1 stick) butter or
  margarine, melted
3 tablespoons lemon juice

Grated rind of 6 lemons
1 cup grated Parmesan cheese
Freshly ground black pepper to
  taste

Simmer the cream over low heat 10 minutes; do not boil. Cook the pasta according to package directions. Drain and return to the pot. Over low heat, stir in the butter, lemon juice, lemon rind, cheese, and warm cream. Mix thoroughly. Add the pepper.

Yield: 6 servings.

Lemon Risotto

3 tablespoons olive oil
¼ cup finely minced onion
1½ cups Arborio rice, uncooked
⅔ cup lemon juice
5 cups chicken broth

1¼ cups fresh peas, blanched, or frozen peas (thawed but uncooked)
¼ cup light cream
⅓ cup grated Parmesan cheese

Heat the olive oil in a heavy 4-quart casserole over moderate heat. Add the onion and sauté 2 minutes to soften, but do not brown. Add the rice to the sautéed onion and stir until all the grains are coated. Stir in the lemon juice until completely absorbed. Begin to add the broth, ½ cup at a time, stirring constantly until all the broth is absorbed into the rice (the rice should be moist but not watery). This should take 18 to 20 minutes, leaving the rice tender but still firm.

As soon as all the broth is absorbed, add the peas, cream, and cheese, stirring constantly until combined with the rice, which should take no more than 5 minutes. Serve immediately.

Yield: 6 servings.

Braised Endive

8   heads endive, washed and
    dried
2   tablespoons butter
Pinch of salt

½   teaspoon sugar
¼   cup lemon juice
¼   cup water

Trim ½ inch from the bottom of each endive. Melt the butter in a sauce-pan or frying pan large enough to hold a single layer of endives. Place the endives flat and add all the remaining ingredients.

Cover the pan and cook over medium heat 15 minutes; turn the endives and continue cooking 10 more minutes, or until all are golden brown. Serve hot.

Yield: 8 servings.

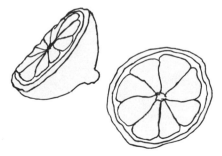

Main Courses

Chicken Teriyaki

3 chicken breasts, split with bone in

¾ cup lemon juice

3 tablespoons soy sauce

½ teaspoon finely grated fresh ginger

1 clove garlic, peeled and minced

½ cup olive oil

Combine all the ingredients; marinate the chicken pieces overnight, in a covered container in the refrigerator. Barbecue on a grill or broil 20 to 30 minutes, turning once and basting frequently with the marinade.

Yield: 6 servings.

Chicken with Lemon Cream Sauce

---

1 chicken, 3 to 4 pounds, cut
  into serving pieces
2 tablespoons flour
⅛ teaspoon salt
Freshly ground black pepper to
  taste
¼ cup (½ stick) butter or
  margarine

½ cup chopped onion
1 pound fresh mushrooms,
  thinly sliced
½ cup lemon juice
Grated rind of 2 lemons
2 cups light cream

Preheat the oven to 350° F.

Rinse the chicken and pat dry. Dredge the pieces in the flour, seasoned with the salt and pepper. Melt the butter in a large skillet over medium-high heat and brown the chicken in batches on all sides. Lower the heat and cook, covered, 20 minutes. Place the chicken in a baking dish.

In the same skillet, sauté the onions and mushrooms 5 minutes. Add the lemon juice and lemon rind, being careful to dissolve all the browned bits left in the pan from the chicken. Add the cream slowly, stirring constantly. When blended, pour over the chicken and bake 30 minutes.

Yield: 4 to 6 servings.

Chicken with Lemon and Rosemary

1 young chicken, 2½ to 3 pounds, or 2 Rock Cornish hens, 1½ pounds each

3 tablespoons olive oil

2 lemons, cut into thin wedges

2 teaspoons chopped fresh rosemary or 1 teaspoon dried rosemary

Preheat the oven to 350° F.

Rub the chicken with the oil and place in a shallow baking pan. Pour any leftover oil into the pan. Place 3 or 4 lemon wedges inside the chicken, along with a pinch of rosemary. Spread the remaining lemon wedges in the pan around the chicken, and sprinkle the rest of the rosemary over the chicken.

Cover the pan with foil and bake 50 minutes. Remove the foil, baste the chicken with drippings, and bake 20 minutes more.

Yield: 4 servings.

Lemon Chicken with Ginger

1 chicken, 3 to 4 pounds, cut into serving pieces

1 teaspoon finely grated fresh ginger

8 ounces apricot jam

½ cup lemon juice

¼ cup (½ stick) butter or margarine

Preheat the oven to 275° F.

Place the chicken pieces in a baking dish. Mix all the other ingredients in a saucepan over medium heat, stirring until the butter melts and the sauce is smooth. Pour the sauce over the chicken and bake, uncovered, 2½ hours.

Yield: 4 to 6 servings.

Chicken with Mushrooms, Olives, and Herbs

½ cup (1 stick) butter, melted
4 large chicken breasts, boned and split (8 pieces)
1 pound fresh mushrooms, thinly sliced
1 cup pitted fresh black olives
3 tablespoons capers
½ cup chopped fresh parsley
2 teaspoons chopped fresh oregano or 1 teaspoon dried oregano
2 teaspoons chopped fresh basil or 1 teaspoon dried basil
4 scallions, chopped
¼ cup olive oil
⅓ cup lemon juice
Grated rind of 1 lemon
⅓ cup dry white wine

Pour the butter over the bottom of an 8-by-12-inch pan or lasagna baking dish. Place the chicken pieces in a single layer in the pan and sprinkle all the remaining ingredients on top. Cover the pan with plastic wrap and refrigerate at least 6 hours. When ready to cook, preheat the oven to 350° F., remove the plastic wrap, and bake 40 minutes.

Yield: 8 servings.

Chicken Fricassee

1 chicken, 3 to 3½ pounds, cut into serving pieces
3 cups water
1 teaspoon salt
1 medium onion, coarsely chopped
Tops of 2 stalks celery
6 whole black peppercorns
5 whole allspice berries
4 carrots, peeled and coarsely chopped
4 stalks celery, chopped

3 tablespoons butter or margarine
¼ cup flour
1 egg yolk, slightly beaten
½ cup light cream
2 tablespoons lemon juice
Salt to taste
½ teaspoon sugar
¼ teaspoon freshly ground black pepper
1 lemon, sliced, for garnish
2 tablespoons chopped fresh parsley for garnish

In a large pot, covered, simmer the chicken in the water with the salt, onion, celery tops, peppercorns, and allspice berries 45 minutes. Add the chopped carrots and celery and simmer 10 minutes more. Remove the chicken, carrots, and celery from the pot and keep on a heated platter in a low oven (200° F.). Strain the broth, discarding the solids, and return the liquid to the pot.

Melt the butter in the top of a double boiler; stir in the flour until smooth and add 2 cups hot broth. Cook the sauce over boiling water until thick, stirring constantly. Combine the egg, cream, and lemon juice and add to the sauce. Continue stirring 1 minute and season with the salt. Add the sugar and pepper.

Pour the sauce over the chicken and garnish with the lemon slices and parsley. Serve with rice.

Yield: 4 to 6 servings.

Veal Piccata

1¼ pounds veal, thinly sliced as for scaloppine
3 tablespoons flour
¼ cup (½ stick) butter
8 slices prosciutto, chopped
¼ cup water
¼ cup lemon juice
2 tablespoons chopped fresh parsley

Flatten the veal slices by pounding them with a mallet between pieces of waxed paper. Dredge the meat in the flour, shaking off the excess. Heat a large nonstick skillet over high heat, add the butter, and, when melted, brown the veal, about 1 minute on each side. Reduce the heat to low and add the water. Place the prosciutto on top of the veal and simmer, uncovered, 20 minutes.

Remove the veal to a warm platter. Pour the lemon juice into the pan and, over medium heat, blend it well with the drippings left from the veal. Pour this sauce over the veal, sprinkle on the parsley, and serve. The dish can also be made with boneless chicken or turkey breasts, although cooking time may vary slightly.

Yield: 4 servings.

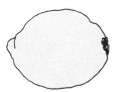

Osso Buco

10 PIECES OSSO BUCO (ASK THE BUTCHER TO PREPARE
LARGE, 3-INCH VEAL BONES WITH MARROW)

⅓ cup flour

¼ cup (½ stick) butter or
   margarine

3 tablespoons olive oil

¾ cup dry white wine

4 large lemons, cut into wedges

2 cloves garlic, peeled and
   chopped

2 cups chicken broth

2 cups beef broth

3 teaspoons chopped fresh basil
   or 1 teaspoon dried basil

3 teaspoons chopped fresh
   rosemary or 1 teaspoon dried
   rosemary

2 medium onions, chopped

4 stalks celery, chopped

4 large carrots, peeled and
   chopped

1 cup chopped fresh parsley
   (½ cup for garnish)

Grated rind of 4 lemons for
   garnish

Shake the meat and flour in a paper bag until evenly coated. Brown the meat on all sides in a frying pan, a few bones at a time, in the butter and oil. Put the browned meat in a large ovenproof pot. Pour the wine into the same frying pan and use it to deglaze the bottom of the pan, then add it to the pot along with all the other ingredients, except the lemon rind and ½ cup of the parsley.

Cover the pot and cook in a 300° F. oven 4 hours. Cool and refrigerate if not serving immediately. Osso buco can be made several days in advance and reheated in a 300° F. preheated oven.

When serving, offer demitasse spoons for extracting the marrow from the bone. This dish is usually served with rice for the broth and toasted bread slices to be spread with the marrow. Serve the lemon rind and remaining ½ cup parsley in separate dishes as garnish for the osso buco.

Yield: 10 servings.

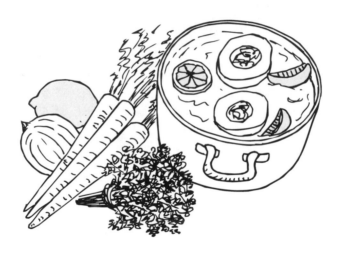

Pork Roast

2½ to 3 pounds boneless loin of pork
2 teaspoons chopped fresh rosemary or 1 teaspoon dried rosemary
⅓ cup lemon juice
⅓ cup sugar

Preheat the oven to 350° F.

Sprinkle the pork with a pinch of the rosemary and place the meat in a roasting pan. Cover tightly with aluminum foil. Roast 1½ hours. Remove the foil and pour off the fat. Return the uncovered roast to the oven for ½ hour, basting often with a mixture of the lemon juice, sugar, and the remaining rosemary.

Yield: 4 servings.

Baked Lemon Swordfish

Thin slices of lemon (enough to cover bottom of pan)

1-inch-thick swordfish or salmon steaks or thick bluefish fillets, ½ to ¾ pound per person

2 teaspoons mayonnaise (see Lemon Notes) for each portion

1 teaspoon capers for each portion

Preheat the oven to 400° F.

Cover the bottom of a broiling pan with the lemon slices and place the fish on top. Coat the tops of the fish with the mayonnaise and capers. Cover the pan with foil and bake 30 minutes. Uncover and brown under the broiler.

Crumb-Baked Flounder

2 pounds flounder fillets, ½ inch thick
½ cup bread crumbs
2 tablespoons butter or margarine, melted
2 tablespoons olive oil
⅓ cup lemon juice
1 tablespoon chopped fresh parsley
¼ teaspoon paprika
⅛ teaspoon freshly ground black pepper

Preheat the oven to 350° F.

Place the fillets in one layer in a baking dish. Combine the remaining ingredients and spread the mixture evenly over the fish. Bake, uncovered, 20 minutes, or until the fish flakes easily.

Yield: 4 servings.

Baked Shrimp

1 pound uncooked shrimp, shelled and deveined
½ cup olive oil
¼ cup lemon juice
Grated rind of 1 lemon

2 tablespoons chopped fresh parsley
1 tablespoon honey
2 tablespoons soy sauce
Pinch of cayenne

Combine the shrimp with the remaining ingredients in a baking dish 9 by 13 inches, stirring to make sure the shrimp is coated. Cover and marinate in the refrigerator at least 2 hours.

Preheat the oven to 450° F. Bake, uncovered, 10 minutes, stirring once. Serve with rice.

Yield: 4 servings.

Avocado-Crabmeat Gratin

2　tablespoons butter
½　pound fresh mushrooms,
　　thinly sliced
⅔　cup chicken broth
1　tablespoon flour
1　pound fresh crabmeat

1　avocado, pitted, peeled, and
　　thinly sliced lengthwise
2　tablespoons lemon juice
1　tablespoon bread crumbs
2　teaspoons grated lemon rind

Preheat the oven to 325° F.

Heat 1 tablespoon butter in a saucepan over high heat and add the mushrooms. Sauté the mushrooms until they give up their liquid, about 5 minutes, and add the chicken broth. Simmer over very low heat.

In another pan, over low heat, melt the remaining tablespoon of butter and blend in the flour with a wooden spoon. When the butter and flour are well combined, add the mixture, a little at a time, to the mushrooms and broth. Increase the heat to high and bring the sauce to a boil, stirring constantly. Remove from the heat.

Arrange the crabmeat and the avocado slices in alternating layers in a 1½-quart casserole dish. Sprinkle each layer with the lemon juice and pour on the mushroom sauce. Top with the bread crumbs and lemon rind. Bake, uncovered, 20 minutes.

Yield: 4 servings.

Shrimp and Bean Salad

1 pound uncooked medium
  shrimp, shelled and deveined
Salt to taste
1 pound string beans
6 tablespoons olive oil
2 tablespoons lemon juice
1 tablespoon white wine vinegar
1 clove garlic
Fresh herbs, 1¼ teaspoons of
  three, and preferably all, of
  the following, chopped: basil,
  thyme, tarragon, oregano, dill

Freshly ground black pepper to
  taste
2 tablespoons chopped fresh
  parsley

Drop the shrimp into 2 quarts of salted, boiling water and cook 1 minute. Drain the shrimp in a colander, allow to cool, then put into a covered container and chill in the refrigerator 2 hours. Snap off the ends of the beans and remove strings, but do not cut the beans; put them in a saucepan with just enough water to cover, and cook until the water begins to boil. Remove the beans immediately from the heat and plunge into cold water to stop from cooking any further. Drain the beans and wrap in paper towels. Chill at least 2 hours in the refrigerator.

To make the dressing, process the remaining ingredients, except the parsley, in a blender or food processor until smooth.

When ready to serve, put the shrimp and beans in a large bowl and combine with the dressing. Toss gently until evenly coated. Sprinkle with the parsley.

Yield: 4 servings.

Sauces and Dressings

Hollandaise Sauce

4 large egg yolks	½ cup (1 stick) butter or
2 tablespoons lemon juice	margarine, softened
¼ teaspoon salt	½ cup boiling water

Put the first four ingredients into a blender, cover, and blend on high speed until thoroughly mixed. Turn to low speed and add the water. The sauce may be reheated in a double boiler, but do not allow the water in the bottom pan to touch the pan holding the sauce.

Yield: 2 cups.

HOLLANDAISE SAUCE BY HAND

In a double boiler, over simmering water, thoroughly combine 4 large egg yolks, 2 tablespoons lemon juice, and ¼ teaspoon salt, using a whisk. Add ½ cup (1 stick) softened butter gradually, stirring constantly. If the sauce is too thick, add 1 tablespoon boiling water.

Mayonnaise

THIS RECIPE MAY BE MADE USING A BLENDER,
A FOOD PROCESSOR, OR A WIRE WHISK.

1	egg	2	tablespoons lemon juice
¼	teaspoon powdered mustard	¾	cup vegetable or olive oil (or
¼	teaspoon salt		a combination of the two)

Blend together the first three ingredients. While continuing to whisk or while the machine is still running, slowly add the lemon juice. When the juice has been incorporated, and while you are still blending the mixture, slowly add the oil until the mixture is thoroughly emulsified.

Cover and store in the refrigerator.

Yield: 1 cup.

Lemon Vinaigrette Dressing

½ cup lemon juice
2 tablespoons chopped fresh parsley
2 teaspoons finely chopped onion

2 teaspoons capers
½ cup olive oil
Salt and freshly ground black pepper to taste

Whisk together the lemon juice, parsley, onion, and capers. Gradually add the olive oil, beating well until a thick consistency has been achieved. Add the salt and pepper.

Yield: 1½ cups.

Bittersweet Salad Dressing

½   cup lemon juice   ⅓   cup olive oil
½   cup honey

Whisk together the lemon juice and honey until the honey has dissolved. Gradually add the olive oil, beating well until a thick consistency has been achieved. Serve over fruit salad or avocados.

Yield: 1⅓ cups.

Mustard Dressing

2   tablespoons lemon juice   1   tablespoon water
1   teaspoon Dijon mustard   1   tablespoon olive oil

Whisk together the lemon juice, mustard, and water. Add the olive oil, beating well until thick.

Yield: ¼ cup.

Seafood Salad Dressing

3 teaspoons lemon juice
Grated rind of 1 lemon
1 cup sour cream or plain
   yogurt
2 tablespoons mayonnaise (see
   Lemon Notes)

¼ teaspoon dry mustard
4 sprigs fresh dill or
   1 tablespoon dried dill
Dash cayenne

Combine all the ingredients in a blender and mix until smooth. Chill in the refrigerator at least 2 hours. Serve with shrimp, crabmeat, or fresh tuna.

Yield: 1¼ cups.

Lemon Sauce for Ham

1 cup red currant jelly
1 cup lemon juice
1 tablespoon dark brown sugar

3 teaspoons brown or Dijon mustard

Combine all the ingredients, except the mustard, in a small saucepan and bring to a boil, stirring constantly until thickened. Remove from the heat. Cool, add the mustard, and stir until smooth. Serve over cold or warm ham.

Yield: 2¼ cups.

Desserts

Lemon~Cranberry Ice

1   quart (4 cups) fresh
    cranberries
2½ cups water
2   cups sugar

⅓   cup lemon juice
Grated rind of 1 lemon
2   cups cold water

Cook the cranberries in 2½ cups water 10 minutes, until the skins are broken. Strain through a fine sieve to eliminate skin and pulp. Combine the liquid with the remaining ingredients and pour the mixture into a 2-quart glass serving bowl. Freeze until firm, 2 to 3 hours, stirring twice during freezing. Serve in avocado halves as an appetizer or alone as a dessert.

Yield: 8 servings.

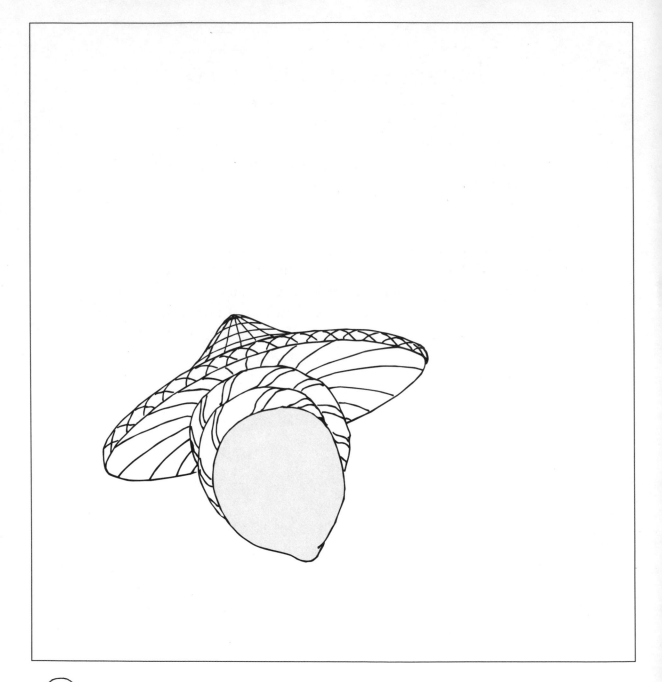

Peach Purée

Approximately 6 fresh peaches,
 peeled and chopped very fine,
 to make 2 cups
1 tablespoon confectioners' sugar

2 tablespoons lemon juice
¼ cup honey
1 teaspoon rum (optional)

Combine all the ingredients in a saucepan and simmer, uncovered, 20 minutes, or until the peaches are tender. Remove from the heat and mash with a fork or a potato masher until smooth. Pour over ice cream or sorbet, or serve over sliced bananas.

Yield: 6 servings.

Jelled Wine Dessert

1½	envelopes unflavored gelatin	¾	cup sugar
½	cup cold water	¾	cup lemon juice
1	cup boiling water	½	cup heavy cream
1	cup white port		

Sprinkle the gelatin into the cold water, stir, and allow to sit 5 minutes. Add the boiling water to the gelatin mixture, then add all the other ingredients except the cream. Stir until the sugar is dissolved.

Pour into individual ramekins or goblets and refrigerate overnight. When ready to serve, whip the cream and add a dollop on top of each serving.

Yield: 6 servings.

Lemon Curd

½ cup lemon juice
Grated rind of two lemons
 1 cup sugar

½ cup (1 stick) butter
 3 large eggs

In a saucepan, over medium heat, combine the lemon juice, rind, sugar, and butter. Stir until the mixture thickens slightly, but do not allow it to boil. Remove from the heat and cool 15 minutes.

In a bowl, using an electric or rotary mixer, beat the eggs until creamy. Add the lemon mixture to the eggs and continue beating until well mixed. Return the mixture to the saucepan and, over low heat, stir until it thickens again, but do not let it boil.

Lemon curd, an English staple, can be spread on toast, muffins, or cookies, or it can be used as a filling for cakes and tarts. If used for tarts, fill baked tart shells (see page 59) and refrigerate until ready to serve. (Tightly covered, lemon curd can be stored in the refrigerator about 2 weeks.)

Yield: 2 cups.

Lemon Bread Pudding

2 large eggs	2 cups milk
¼ cup (½ stick) butter or margarine, softened	2 cups white bread crumbs
1 cup sugar	3 tablespoons lemon juice
	Grated rind of 1 lemon

Preheat the oven to 325° F.

Using a rotary or electric mixer, beat the eggs, butter, and sugar until creamy. Combine the milk and bread crumbs and add to the egg mixture, beating together well. Stir in the lemon juice and rind and pour into a 1-quart baking dish. Place the dish in a pan of hot water and bake, uncovered, 30 minutes.

Yield: 6 servings.

Lemon Bread

1 cup sugar
½ cup (1 stick) butter
2 large eggs, beaten
½ cup milk
1½ cups all-purpose flour

1 teaspoon baking powder
Grated rind of 2 lemons
¾ cup finely chopped walnuts (optional)

Preheat the oven to 350° F.

With a rotary or electric mixer, cream together the sugar and butter. Add the eggs and milk and blend thoroughly. Fold in the flour and baking powder, and, when all are well combined, add the lemon rind and nuts.

Pour into 3 greased and floured loaf pans 3 by 6 inches, and bake 45 minutes. Cool 5 minutes before removing from the pans; pierce the tops of the loaves a few times with a toothpick to speed up the cooling process. If you wish, pour over the tops a mixture of ½ cup sugar dissolved in 3 tablespoons lemon juice. The loaves may be frozen.

Yield: 4 servings per loaf.

Crust for
Pies and Tarts

2¼ cups all-purpose flour	5 tablespoons cold water
1 teaspoon salt	¾ cup Crisco shortening

Sift together the flour and salt in a large mixing bowl. Transfer ½ cup of the flour mixture to another bowl, add the water, and stir into a smooth paste with a rubber spatula.

Add the shortening to the dry flour and salt mixture and, using 2 knives or a pastry blender, cut into the mixture until there is dough the size of small peas.

Stir the smooth flour paste into the dough, working with your fingers or a fork, until it is all well mixed. Use your hands to form the resulting dough into a ball.

When the dough holds together, separate it into two round, flat masses, ready for rolling into crusts. (If only one crust is needed at this time, the other may be wrapped in plastic and stored in the freezer for future use.)

Sprinkle flour on a board, using as little flour as possible to prevent sticking, or roll the dough between sheets of waxed paper. Roll the dough in one direction until it is ⅛ inch thick and 10 inches in diameter. Fold the dough in half, carefully lift it from the board, and unfold it onto a 9-inch pie pan. Prick the bottom of the dough in several places with a fork to avoid shrinkage. Press down the edges of the crust along the rim of the pan with a fork to eliminate excess crust as well as to help the dough adhere to the pan.

If the piecrust needs to be baked before filling, preheat the oven to 350° F. and bake 10 minutes.

Yield: Two 9-inch crusts.

TART CUPS

1   8-cup muffin tin   Dough for one 9-inch piecrust (see
                                                 preceding recipe)

Divide the dough into 8 small balls and flatten for rolling. Turn the muffin tin upside down. Roll out the dough into 2½-inch circles and place the rolled-out rounds on the inverted muffin tin. Mold the edges of each round onto the sides to form a cup. With a fork, prick the bottom of each crust.

Preheat the oven to 350° F. Place the muffin tin, in its upside-down position, on a cookie sheet and bake 8 minutes. Cool before removing the pastry cups from the tin.

Yield: 8 cups.

Grandma's Lemon Meringue Pie

1 cup plus 6 tablespoons sugar
2 tablespoons flour
3 tablespoons cornstarch
Pinch of salt
1 cup boiling water
3 large eggs, separated

½ cup lemon juice
Grated rind of 1 lemon
2 tablespoons butter
1 baked cooled 9-inch piecrust
(see page 58)

Preheat the oven to 350° F.

In a saucepan, over medium heat, combine 1 cup of the sugar with the flour, cornstarch, and salt. Stir in the water and continue stirring until the mixture thickens. Remove the saucepan from the heat and, when the mixture is cool, add to it the egg yolks, slightly beaten, lemon juice, lemon rind, and butter. Return the saucepan to the heat. Stir until the custard thickens again. Pour it into the piecrust.

Make the meringue by beating the egg whites, with a rotary or electric mixer, until frothy. Continue to beat while gradually adding the 6 remaining tablespoons sugar. Keep beating until stiff peaks form. Spread the meringue over the pie, making sure it touches the edges of the crust all around (to prevent it from shrinking). Bake 15 minutes.

Yield: 8 servings.

Fluff Pie or Tarts

4 large egg yolks
⅔ cup plus 5 tablespoons sugar
¼ cup lemon juice
Grated rind of 1 lemon
4 large egg whites, divided into
   2 bowls

1 baked, cooled 9-inch piecrust
   or 8 baked tart cups (see page
   59)

Beat the egg yolks until creamy, using a rotary or electric mixer. Add ⅔ cup of the sugar and continue beating. Add the lemon juice and rind and, when well mixed, pour into the top of a double boiler. Cook over boiling water 12 minutes, stirring constantly. Remove from the heat and let cool.

Preheat oven to 300° F. Beat 2 of the egg whites (see NOTE, p. 63) until frothy. Add 1 tablespoon of the sugar and continue beating until stiff. Fold into the cooled lemon mixture and pour all into the piecrust or tart cups.

Beat the remaining 2 egg whites with the last 4 tablespoons sugar. When the mixture is stiff and glossy, spread over the pie, taking care to cover all the edges of the crust with the meringue to prevent shrinkage. Bake 15 minutes.

Yield: 8 servings.

NOTE: Wash the mixer carefully after beating the egg yolks. Just a drop of yolk in egg whites may prevent the whites from becoming stiff when beaten. For best results, egg whites should be at room temperature.

Chess Pie

2 cups sugar
½ cup (1 stick) butter or margarine, softened
4 large eggs
1 tablespoon cornstarch

½ cup lemon juice
Grated rind of 2 lemons
1 unbaked 9-inch piecrust (see page 58)

Preheat the oven to 350° F.

Cream the sugar and butter with a rotary or electric mixer. Add the eggs, one by one, beating well after each addition. Add the cornstarch, lemon juice, and rind and mix well.

Pour the filling into the piecrust and bake 45 minutes. Cool before serving.

Yield: 8 servings.

Angel Pie

MERINGUE

4 large eggs whites	1 cup sugar
¼ teaspoon cream of tartar	

Preheat the oven to 275° F. Place a piece of heavy brown paper on a cookie sheet. Using a pencil, draw a 9-inch circle on the paper.

Beat the egg whites with the cream of tartar until frothy. Keep beating and gradually add the sugar until the whites form stiff peaks.

Spread the meringue mixture evenly on the brown paper, covering all of the 9-inch circle, and bake 1 hour. Turn off the oven, set the door ajar, and leave the meringue inside the oven until cool. When the meringue has cooled, transfer it carefully from the brown paper to a serving plate.

LEMON FILLING

4 large egg yolks	⅓ cup lemon juice
½ cup sugar	Grated rind of 1 lemon

Beat the egg yolks until creamy. Add the sugar and continue beating while adding the lemon juice and rind. Pour the mixture into the top of a double boiler over boiling water and cook until thick, stirring constantly. Cool, then spread the lemon custard over the entire meringue.

TOPPING

1   cup heavy cream, whipped

Spread the cream on top of the pie. Refrigerate overnight. Cut into pie-shaped wedges to serve.

Yield: 8 to 10 servings.

Lemon Mousse

6   large eggs, separated
1   cup plus 2 tablespoons sugar
1   cup lemon juice

3   tablespoons grated lemon rind
1   pint (2 cups) heavy cream

Using a rotary or electric mixer, beat the egg yolks until creamy, gradually adding 1 cup of the sugar. In a double boiler, over simmering water, add yolk mixture and stir with a wire whisk until it thickens into a custard. Remove from heat and cool. Slowly add the lemon juice and rind while continuing to beat.

In another bowl, beat the egg whites (see NOTE below) until stiff, gradually adding 1 tablespoon of the sugar. Whip the cream in a third bowl, adding the remaining sugar while whipping.

Fold all mixtures together and pour into a large glass bowl. Refrigerate overnight.

Yield: 8 servings.

NOTE: Wash the mixer carefully after beating the egg yolks. Just a drop of yolk in egg whites may prevent the whites from becoming stiff when beaten. For best results, egg whites should be at room temperature.

Lemon Soufflé

6 large eggs, separated	¼ teaspoon salt
1 cup sugar	Grated rind of 1 lemon
4 tablespoons lemon juice	

Preheat the oven to 325° F.

With a rotary or electric mixer, beat the egg yolks until light and creamy. Add the sugar and lemon juice, beating all the time.

In a separate bowl, beat the egg whites (see NOTE below) with the salt until stiff but not dry.

Fold the egg whites into the yolks, and pour into a 6-cup soufflé dish. Sprinkle the top with the lemon rind. Place the dish in a pan of hot water and bake 50 minutes. Serve at once.

Yield: 6 to 8 servings.

NOTE: Wash the mixer carefully after beating the egg yolks. Just a drop of yolk in egg whites may prevent the whites from becoming stiff when beaten. For best results, egg whites should be at room temperature.

Apricot Soufflé with Custard Sauce

11 ounces dried apricots
1½ cups water
1¾ cups sugar
½ cup lemon juice

Grated rind of 1 lemon
6 large eggs, separated
1½ cups milk
1 teaspoon vanilla extract

Preheat the oven to 300° F.

Cook the apricots in the water until soft, about 15 minutes, and mash them while hot with a potato masher. Add 1½ cups of the sugar, the lemon juice, and the rind.

Beat the egg whites until stiff, with a rotary or electric mixer, and fold them into the apricot mixture. Pour into a 2-quart soufflé dish, set the dish in a pan of hot water, and bake 30 minutes. Turn the oven off and set the door ajar, allowing the soufflé to cool in the oven.

To make the custard for the topping, beat the egg yolks until creamy, adding the milk and remaining 4 tablespoons sugar. Cook in the top of a double boiler over boiling water, stirring with a wooden spoon until the custard is thick enough to coat the spoon. Remove from the heat, add the vanilla extract, cover, and store in the refrigerator until ready to serve.

The soufflé should be served the day it is made, at room temperature. Spoon the sauce, chilled, over individual portions.

Yield: 8 servings.

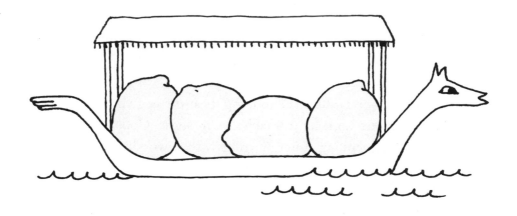

Apple Crisp

⅓ cup all-purpose flour
⅔ cup nonfat dry milk
⅔ cup sugar
2 teaspoons ground cinnamon
½ cup (1 stick) butter or
    margarine, softened

6 to 8 large tart apples, peeled,
    cored, and thickly sliced
2 tablespoons lemon juice
Grated rind of 1 lemon

Preheat the oven to 325° F.

With a fork, mix together the flour, dry milk, ⅓ cup sugar, and 1 teaspoon cinnamon. Add the butter and continue to mix until crumbly. In another bowl, combine the apples with the remaining sugar, 1 teaspoon cinnamon, and the lemon juice and rind; pour into a greased shallow Pyrex dish, 9 by 13 inches, or a 10-inch pie dish.

Sprinkle the crumble mixture over the apples and bake 45 minutes, or until the apples are tender and the topping is crisp. Serve warm with cream or vanilla ice cream. (One pound blueberries can be used in place of the apples.)

Yield: 5 to 6 servings.

Baked Pears

4 pears, peeled, cored, and
   sliced ¼ inch thick
⅓ cup lemon juice

Grated rind of 1 lemon
½ cup dark brown sugar
2 tablespoons honey

Preheat oven to 350° F. Place the pears in a shallow 9-inch pie dish, and pour the lemon juice over them. Sprinkle with the lemon rind and brown sugar and drizzle the honey over the top. Bake 40 minutes, or until soft. Serve with cream or ice cream.

Yield: 4 servings.

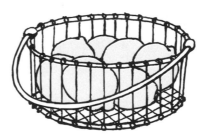

Lemon Pound Cake

1 cup (2 sticks) butter
2 cups sugar
6 large eggs, separated
8 ounces cream cheese, softened

1 tablespoon lemon extract
2 cups cake flour
Grated rind of 2 lemons

Preheat the oven to 350° F.

Cream the butter and sugar thoroughly with a rotary or electric mixer. Add the egg yolks, one at a time, while continuing to beat. Add the cream cheese and lemon extract, beating until light and fluffy. Add the flour gradually, and beat until the mixture is smooth. Add the lemon rind.

Beat the egg whites (see NOTE below) until stiff, and fold into the lemon mixture. Pour into a greased loaf pan 9 by 5 by 3 inches. Bake 45 minutes.

Yield: 8 servings.

NOTE: Wash the mixer carefully after beating the egg yolks. Just a drop of yolk in egg whites may prevent the whites from becoming stiff when beaten. For best results, egg whites should be at room temperature.

Sponge Cake

3	large eggs, separated	2	tablespoons butter or	
1½	cups milk		margarine, melted	
1	cup sugar	3	tablespoons lemon juice	
3	tablespoons flour	Grated rind of 1 lemon		

Preheat the oven to 350° F. Beat the egg yolks well with a rotary or electric mixer, and add the milk. In another bowl, mix the sugar and flour thoroughly with the butter, lemon juice, and rind. Add to the egg mixture and stir well.

Beat the egg whites (see NOTE below) until stiff and fold them into the egg mixture. Pour into a 1-quart casserole dish. Put the dish into a pan of hot water and bake 40 minutes.

Yield: 6 servings.

NOTE: Wash the mixer carefully after beating the egg yolks. Just a drop of yolk in egg whites may prevent the whites from becoming stiff when beaten. For best results, egg whites should be at room temperature.

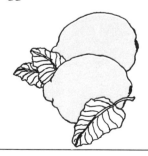

Lemon Icing

4 large egg whites, at room temperature

2 cups sugar

½ cup cold water

4 teaspoons lemon juice

With a rotary or electric mixer, beat the egg whites until frothy. In a double boiler over boiling water, combine the sugar with the cold water and lemon juice. When the mixture "drips" in drops from a teaspoon, it is ready to be beaten slowly into the egg whites with a whisk. Continue beating until firm.

Yield: Enough to cover a large 10-inch cake.

Buttercream Frosting

2 cups confectioners' sugar
½ cup (1 stick) butter, softened

3 tablespoons lemon juice
3 tablespoons milk

With a rotary or an electric mixer, beat together all the ingredients until smooth. Add more milk, a little at a time, if frosting needs thinning for spreading on a cake.

Yield: Enough to cover a 2-layer cake or 24 cupcakes.

Lemon Velvet Ice Cream

5 cups heavy cream	2 teaspoons lemon extract
5 cups milk	4 cups sugar
2 cups lemon juice	Grated rind of 2 lemons

Mix all the ingredients thoroughly with a wooden spoon, and pour into a 4-quart container. Put in the freezer overnight. Let soften 20 minutes before serving.

Yield: 30 ½-cup servings.

Parfait-in-a-Lemon

6   large lemons
6   scoops vanilla ice cream

½   cup fresh or frozen raspberries,
     mashed with a fork

Cut a tiny slice off one end of each lemon, just enough so that the lemon will stand up on a plate.

Cut a 1-inch slice off the other end, which will be the end for scooping out the fruit and putting in the ice cream. With a melon baller or a grapefruit spoon, scoop out the pulp of the lemon and some of the pith. Do this over a bowl to catch all the insides, including the juice. Discard the pith and pour 2 tablespoons of the fruit and juice back into each lemon.

Put a scoop of ice cream into each lemon, and place the lemons in the freezer at least 1 hour. When ready to serve, top each lemon with 1 tablespoon of the raspberries. An excellent variation is lemon sorbet with crème de menthe topping.

Yield: 6 servings.

Lemon Squares

BOTTOM LAYER

1 cup all-purpose flour	¼ cup butter, melted
¼ cup confectioners' sugar	

Preheat the oven to 350° F.

Mix the ingredients together with a fork, and press the batter down with hands to cover the bottom of a metal pan 8 by 8 inches. Bake 15 - *19* minutes. Cool slightly.

TOP LAYER

2 large eggs, beaten well	⅓ cup lemon juice
1 cup sugar	Grated rind of 1 lemon
2 tablespoons flour	¼ cup confectioners' sugar
½ teaspoon baking powder	

With a rotary or an electric mixer, combine all the ingredients well, except for the confectioners' sugar, and pour the mixture over the bottom crust. Bake 20 to 25 minutes. Cool.

Sprinkle with the powdered sugar and cut into squares.

Yield: 30 squares.

Lemon Tea Cookies

½	cup granulated sugar	4	tablespoons lemon juice
½	cup confectioners' sugar	2	tablespoons grated lemon rind
½	cup (1 stick) butter, softened	2	cups all-purpose flour
2	tablespoons Crisco shortening	½	teaspoon baking soda
1	large egg		Powdered cocoa

In a large bowl, using a rotary or an electric mixer, beat the sugars, butter, and shortening until well combined. Add the egg, lemon juice, and rind, and beat until light and fluffy. Add the flour and baking soda and mix well.

Cover the bowl air tight with plastic wrap and refrigerate at least 2 hours so the dough can become solid. (The dough may be stored in the freezer until ready for use.)

When ready to bake, preheat the oven to 350° F. With hands, shape the chilled dough into 1-inch balls, and place the balls 2 inches apart on an ungreased cookie sheet.

Bake 10 minutes and remove the cookies immediately from the sheet. Cool 5 minutes, then lightly dust the tops of the cookies with the cocoa.

Yield: 30 cookies.

Lemon Wafers

⅔ cup sugar
½ cup (1 stick) butter, softened
1 egg yolk

2 tablespoons lemon juice
¾ cup flour
Grated rind of 1 lemon

Cream the sugar and butter with a rotary or an electric mixer. Beat in the egg yolk until smooth. Add all the other ingredients, beat well, cover, and chill in the refrigerator overnight.

When ready to bake, preheat the oven to 350° F. Drop the dough 1 inch apart with a teaspoon onto an ungreased baking sheet and bake 10 minutes, or until the edges are golden brown. Allow the wafers to cool 1 minute, or until firm, before removing them from the baking sheet.

Yield: 30 cookies.

Oat Thins

½ cup (1 stick) butter, softened
¾ cup light brown sugar
1 egg
¾ cup flour

¼ cup rolled oats
½ cup ground almonds
Grated rind of 1 lemon
3 teaspoons lemon juice

Preheat the oven to 350° F.

Cream the butter and sugar with a rotary or an electric mixer. Add the egg and mix well. Stir in the flour, oats, almonds, lemon rind, and lemon juice, and mix until smooth. Drop the dough with a teaspoon onto a greased cookie sheet, 2 inches apart, and bake 8 to 10 minutes.

Yield: 30 cookies.

Roman Cookies

½   cup (1 stick) butter, softened
⅓   cup light brown sugar
Finely grated rind of two lemons
1   cup all-purpose flour

4   teaspoons lemon juice
2   additional lemons
2   tablespoons granulated sugar

Beat the butter and sugar until creamy with a rotary or an electric mixer. Add the lemon rind, flour, and lemon juice and blend well. Roll the dough into a log shape 1½ inches in diameter and wrap it in a piece of waxed paper. Refrigerate until well chilled. (The dough may be stored in the freezer until ready for use.)

When ready to bake, preheat the oven to 350° F. Grate the rind of the 2 additional lemons and combine with the granulated sugar. Slice the dough log into rounds ¼ inch thick. Place them on an ungreased cookie sheet ½ inch apart. Put a bit of the rind-sugar mix on top of each cookie.

Bake 12 minutes, or until golden and lightly browned at the edges. Remove from the oven and cool on a cookie rack.

Yield: 30 cookies.

Lemon Meringue Kisses

2   egg whites, at room
    temperature
⅔   cup sugar

3   tablespoons lemon juice
    Grated rind of 1 lemon

Preheat the oven to 275° F.

Beat the egg whites until stiff. Add the sugar and continue beating until glossy. Still beating, slowly add the lemon juice and rind. Line a cookie sheet with brown paper and, using a teaspoon, drop dollops of the meringue onto the paper about 1 inch apart. Bake 40 minutes, turn the oven off, open the door slightly, and leave it ajar until the meringues cool.

Yield: 40 to 50 kisses.

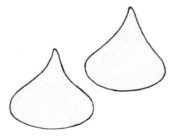

Beverages

Strawberry Punch

1   cup fresh or frozen
    strawberries, crushed
½   cup sugar (if using frozen
    strawberries, sweeten to taste)

½   cup lemon juice
2   cups water
1   quart ginger ale

Mix all the ingredients and serve over ice.

Yield: 16 servings.

Iced Tea

3 cups boiling water
4 teaspoons loose tea or 4 tea bags
½ cup fresh mint leaves
1 cup sugar

4 cups cold water
¾ cup lemon juice
1 cup orange juice
Lemon slices (optional)

Pour the boiling water over the tea, add the mint and sugar, and steep the tea until it cools. Strain, removing the tea and mint. Add the last three ingredients, chill, and serve with lemon slices, if desired.

Yield: 8 to 10 servings.

New Year's Eve Punch

1   fifth brandy
1   fifth white rum
1   fifth dark rum
1   cup peach schnapps

1½   cups sugar
2   quarts green tea
3   cups lemon juice

Stir all the ingredients together with a block of ice in a large punch bowl.

Yield: 40 servings.

Nonalcoholic
New Year's Eve Punch

1 cup grapefruit juice
1 cup lemon juice
3 cups orange juice
1 quart apple juice

10 ounces (1¼ cups) fresh or
frozen strawberries
1 quart ginger ale

Mix together the grapefruit juice, lemon juice, and 3 cups cold water. Pour into ice-cube trays and freeze. Empty the trays into a large punch bowl and add the remaining ingredients, plus 2 cups cold water.

Yield: 20 servings.

Mulled Cider

1   lemon, sliced very thin
½   orange, sliced very thin
2   teaspoons lemon juice
4   3-inch cinnamon sticks

2   tablespoons whole cloves
2   cups dark rum
2   quarts apple cider

Heat all the ingredients in a saucepan, but do not bring to a boil. Pour into mugs and serve warm.

Yield: 6 to 8 servings.

Hot Toddy

2 tablespoons bourbon
1 tablespoon honey

1 tablespoon lemon juice
¾ cup very hot water

Combine all the ingredients and sip the drink slowly while it's hot. Best taken at bedtime, this toddy is very good for bronchial and flu symptoms.

Yield: 1 serving.

Miscellany

Lemon Marmalade

2   large lemons, seeded and
    chopped into small pieces
2   large carrots, peeled and
    coarsely chopped
⅔   cup honey

1¼ cup sugar
½   cup water
½   cup orange juice
⅛   teaspoon grated fresh ginger

Put all the ingredients into a saucepan and cook, covered, over low heat
1 hour. Remove the cover and cook 30 minutes longer, stirring occasionally, until the juice is thick. Cool and put into sterilized jars. Store in the
refrigerator.

Yield: 1 pint.

Lemon Pickle

2	large or 3 small lemons	1	teaspoon cayenne
1	teaspoon salt	1	teaspoon grated fresh ginger
2	cloves garlic, peeled and minced	2	teaspoons powdered mustard
½	cup (3 ounces) golden raisins	½	cup cider vinegar
8	cloves	1	cup sugar

Slice each lemon into 5 or 6 crosswise slices and remove all the seeds. Sprinkle the slices with the salt, and lay the slices in several layers in a flat dish. Cover and refrigerate 3 days. On the fourth day, add to the lemon slices a well-blended mixture of the garlic, raisins, cloves, cayenne, ginger, mustard, and vinegar. Cover the dish again and return it to the refrigerator for one more day. Stir occasionally during these 4 days.

On the fifth day, strain the mixture into a saucepan and discard the cloves. Mince the lemon-raisin mixture in a blender or food processor. Add this to the saucepan, plus the sugar, and bring to a boil. Reduce the heat and simmer, uncovered, 40 minutes, stirring occasionally. When the mixture starts to thicken, remove it from the heat. Cool and refrigerate, in covered containers.

This lemon pickle will keep several months, if refrigerated. Serve it with curries, cold meats, or chicken.

Yield: 1½ to 2 cups.

Lemon Mints

1½   cups sugar   ½   teaspoon lemon extract
 1   cup water

Boil the sugar and water until the mixture just begins to thicken. Add the lemon extract.

    Caution: The mixture can get too hard in the pot if it cooks too long. With a teaspoon, quickly dot the mixture onto waxed paper.

Yield: Approximately 20 thin "mints."

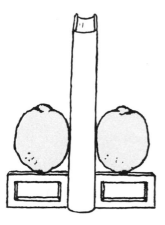

Candied Lemon Peel

16   large lemons   6½   cups sugar

Peel strips ⅓ inch wide and 3 inches long from the lemons, moving length-wise and avoiding the white pith as much as possible. (Use the peeled lemons for juice in other recipes.) Place the strips in a bowl with enough water to cover and soak 2 hours. Drain and cover the peels again with water, this time in a pot, and bring to a boil. Simmer 10 minutes and drain. Cover again with water and boil 30 minutes and drain.

Put 6 cups of water in another pot with 6 cups of the sugar and bring to a boil. Add the lemon peels and stir with a wooden spoon, simmering 45 minutes until they are translucent. Let the peels stand in the syrup overnight.

Drain off the syrup. Place the peels on paper towels and let stand to dry 24 hours. Roll the peels in the remaining ½ cup sugar. Stored in an airtight container, they will keep indefinitely.

Yield: Fills four 6-ounce containers.

Pomander Ball

1   lemon
2   tablespoons whole cloves

Ribbon
Hat pin or long nail

Roll the lemon on a counter, pressing hard to release the juices in order to enhance the aroma. Push the cloves into the lemon (a sharp nail may be necessary to break a tough skin). Tie a bow with the ribbon. Using more ribbon or a cord, make a loop for hanging, and secure it to the top of the lemon with a hat pin or a long nail. Hang in a closet as an air freshener.

Lemon Basket

1   large, firm lemon   Rind peeler or zester
Small paring knife   Bunch of parsley or tiny flowers

Follow the step-by-step drawings and these directions:

1. Use a small paring knife to cut across the bottom of the lemon so it will stand upright.
2. Cut 2 slits halfway down from top center to make the handle.
3. On the lower half of lemon, cut away the rind in strips, with a rind peeler, to make a basket weave.
4. With the paring knife, carefully cut away the top part of the lemon, begun in step 2, leaving the basket handle attached to the bottom and cutting zigzag edges around the lemon.
5. Remove all the pulp from the basket handle.
6. Set the lemon basket on a small porcelain saucer to prevent the citric acid from damaging the table surface.
7. Fill the basket with parsley or tiny flowers by sticking the stems into the pulp left in the bottom of the lemon.

1.

2.

3.

4.

7.

Lemon Tree

Lemon trees can be easily grown as house plants. Place a seed from a fresh lemon in a 2-inch (in diameter) flowerpot filled with moist potting soil, and cover the seed with ¼ inch more of the soil. Enclose the pot in a plastic bag and place it in a warm, shaded spot until it germinates, usually in 2 to 3 weeks.

As soon as germination occurs, remove the plastic bag and place the pot in a sunny place, indoors or out, depending on the climate. The plant is safe outside as long as there is no threat of frost. Fertilize it as you would any other house plant.

Lemon Aides

Our thanks to our editor, Dawn Drzal, and to Fiammetta Ajó, Leigh Butler, Sally Butler, Connie Class, Lucy S. Danziger, Barbara Day, Jane Fiore, Rusty Gelb, Ruth Knight, Ginny Ladd, Paola Lucentini, Karen Marshall, Annie Overholser, Adeline Rand, Shirley Ratterree, Liz Rosenman, Vivian Schulte, Howell Scott, Shirley Sherwood, Bob Silver, Joanne Slusser, Jane Ulstrup, and Frannie Woodruff.

Index

About the Authors

Sarah Schulte is a full-time artist. **Lalitte Scott** is a professional calligrapher with an interest in horticulture. Both women live in New York City, enjoy cooking, and love lemons.